Transmagnetic Resonance Field Theory

By: Timothy E Douglas

A Quantitative Framework on Resonance

Dedication & Acknowledgements

This research project is dedicated to every individual. It is my sincerest hope that this research paves the way for a world where we are no longer having to focus on survival but of ambition and creation. Where every individual can thrive at their own pace without fears or anxieties. Where understanding and compassion replaces ego and condescension.

I'd like to acknowledge the minds of Nikola Tesla, Weiner Heisenburg, Max Planck, Wolfgang Pauli, and Erwin Schrödinger. As they have laid the foundation of what we know of electromagnetism and physics today.

Introduction

Resonance has been a core principle across various fields of physics, from mechanical systems to acoustics, and now, at the cutting edge of quantum science. As our understanding of quantum mechanics deepens, so does our need to explore how quantum resonance impacts both the natural and engineered systems we interact with. This concept extends beyond the classical framework, pushing us into realms where energy transfer, waveforms, and electromagnetic fields behave in ways that challenge traditional laws of physics.

The emergence of transresonance as a quantum phenomenon bridges the gap between familiar electromagnetic interactions and the sophisticated behavior of subatomic particles. At the heart of this exploration is the resonance matrix—a multidimensional field where various forms of energy coalesce, resonate, and exchange properties, leading to phenomena such as quantum leaps and field amplification.

This work builds upon earlier theories of resonance in both classical and quantum domains, presenting a novel hypothesis that positions quantum resonance as an array or matrix, where intersecting points act as quantum nodes for energy exchange. In this context, resonance ceases to be a mere oscillatory motion, instead evolving into a dynamic, interconnected system of intersecting frequencies and quantum states.

Incorporating principles such as the Heisenberg Uncertainty Principle and Planck's constant into this model allows for a more precise, albeit complex, understanding of how resonance fields interact at both macroscopic and quantum levels. The resulting theoretical framework, Transresonance Field Theory (TRFT), posits that resonance is not only a measurable physical property but also the gateway to quantum state transitions, offering implications for advanced technological applications and fundamental quantum research.

This comprehensive investigation introduces new mathematical models to support the hypothesis and details a real-world experimental approach using the transresonator—a device engineered to amplify natural electromagnetic fields through resonance. The synthesis of theoretical and experimental data underscores the potential for TRFT to transform our understanding of quantum systems and opens the door to further exploration into quantum resonance matrices.

Part 1

<u>The Transresonator Device & Construction</u>

Transresonator Construction:

The transresonator device developed for this research is a prototype designed to test the hypothesis that specific resonant frequencies can enhance electromagnetic fields. The device was constructed using a combination of a Bluetooth speaker, an electromagnet, and additional magnet components to form the magnet stack. Below is a breakdown of the core elements and their role in the device:

1. Bluetooth Speaker:

The Bluetooth speaker serves as the base for generating and delivering frequency signals to the transresonator. It was repurposed for its connection to an external signal generator and amplifier setup, allowing the speaker's leads to drive the electromagnet directly.

Amplifier Leads: The internal wiring of the speaker was modified so that the amplifier leads could be connected directly to the electromagnet, bypassing the speaker cone. This allowed the speaker to act as a frequency delivery system without producing sound, purely transmitting the frequency signal into the electromagnetic setup. This method of inducting frequencies I have termed Harmonic Frequency Induction to produce Transmagnetic Resonance Fields.

2. 9V 50N Electromagnet:

A key component of the transresonator, the 9V 50N electromagnet was installed in place of the speaker's diaphragm. The electromagnet was essential for propagating the signal through the electromagnetic field (EMF) that could be influenced by the applied frequencies.

EM Field Properties:

The electromagnet- combined with the speaker cone magnet, connected by a steel bolt, produced an initial field with a strength of approximately 0.42uT (microteslas) when no frequency was applied. Once the 144,000 Hz frequency was introduced, the field strength increased to 0.81uT, demonstrating the influence of frequency on EMF enhancement.

3. Internal Speaker Magnet:

It was observed that a stable magnetic field is needed for the propagation of the signal. Thus, then the internal magnet of the Bluetooth speaker remained, providing a consistent magnetic field that interacted with the newly connected electromagnet and frequency generator. This field, combined with the input from the signal generator, played a role in creating the resonance needed for testing.

4. Magnet Stack:

The magnet stack was added as part of the latest experimental iteration. This stack includes multiple magnets layered to increase the baseline EMF, providing a stronger field to test frequency effects on.

Baseline Field Strength: The magnet stack produced

A base EMF of approximately 4uT, with peaks reaching as high as 18.64uT before any frequencies were applied.

Testing without Power or Signal: When no signal or power was applied, the magnet stack maintained this baseline field, which provided a stable foundation for comparing the effects of frequency input.

5. Signal Generator:

A critical aspect of the experiment involved the use of a signal generator capable of outputting various frequencies, including the solfeggio-based frequencies like 144,000 Hz.

Frequency Range: The generator used in these tests delivered signals within a specified range of harmonic frequencies, focusing on those theorized to be within the 3/6/9 resonance matrix.

Future Testing with Higher Frequencies:

Upcoming tests will involve a spectrum analyzer with a DDS generator capable of generating frequencies up to 350 MHz, allowing for more comprehensive exploration of high-frequency resonance impacts on the electromagnetic fields.

6. EMF Reader:

Throughout the experiment, an EMF reader was used to capture baseline readings and monitor changes in the magnetic field strength when frequencies were applied.

Results:

The EMF readings consistently showed an increase in field strength when resonant frequencies were introduced. For instance, the application of 144,000 Hz caused the field to nearly double in strength, from 0.42uT to 0.81uT. The 3 in 1 device used as a DDS generator, provided Oscilloscope and multi-meter readings as well. The wave pattern on the Oscilloscope matched that of the one sent through the speaker system at 20448Hz.

The field created from my method of Harmonic Frequency Induction is termed a Transmagnetic Resonance Field.

Note: Setup did not interfere with the Bluetooth signal of the speaker. All Frequencies under 22,000Hz were run through the speaker from my cell phone via bluetooth.

7. Potential Modifications and Future

Development:

Spectrum Analyzer: The next phase of testing will include a spectrum analyzer capable of measuring frequencies up to 350 MHz. This will enable exploration of higher-frequency effects on the magnetic field and allow for more precision in identifying resonant frequency thresholds.

Refining the Transresonator Design:

Future iterations of the transresonator will incorporate additional components, such as advanced signal processors and different magnet configurations, to optimize resonance efficiency and magnetic field strength.

<u>Applications of the Transresonator</u>

Main Application:

Resonance Frequency Alignment or Transmagnetic Resonance Therapy

The core function of the transresonator lies in its ability to alter an individual's resonance frequency, affecting both physical and energetic levels. Through the combination of precise electromagnetic fields and sound frequencies, the system induces changes on a subatomic level, causing the user's body and energetic field to resonate at the same frequency as that which is being output by the device. This matching of resonance frequencies could theoretically:

Enhance energetic healing by harmonizing with the body's natural frequencies.

Alter or raise the user's vibrational state, which could have implications for mental clarity, physical wellness, and spiritual ascension.

Serve as a mechanism for accessing altered states of consciousness or higher-dimensional realms.

The effects are achieved by influencing the subtle energy fields that exist within and around the human body. By affecting these fields at a subatomic level, the device allows for potential recalibration of the individual's energetic system to achieve a "tuning" effect, much like how a radio adjusts to receive a specific station. When the frequency applied through the transresonator is finely tuned, the user can theoretically experience both physiological and metaphysical shifts. In short, this device will re-magnetize the crystals in your pineal gland and bring you in tune to whichever frequency is inducted through it.

Resonance vs Magnetism Overview:

This section provides the investigation of the relationship between resonant frequencies and electromagnetic field (EMF) enhancement, demonstrating that specific frequencies can strengthen EM fields. The study introduces the use of a transresonator device, drawing from the theoretical foundation of the 3/6/9 resonance matrix, solfeggio frequencies, and universal resonance scales. Key experimental results provide tangible proof of frequency's influence on EM fields, establishing potential applications in advanced energy manipulation and metaphysical practices.

The hypothesis explored in this investigation is that frequency, particularly resonant frequencies, can significantly enhance or strengthen electromagnetic fields. This work is grounded in theoretical studies of resonance, including the 3/6/9 matrix, solfeggio frequencies, and the emerging field of transresonator technology.

The goal of this research is to offer a tangible understanding of how specific frequencies impact the strength of an EM field and provide a stepping stone towards using resonance as a tool for manipulating Biomagnetic fields. The concept originated from the discovery of resonance matrices within planetary frequencies, solfeggio scales, and the idea that certain frequencies could serve as "error-correcting codes" in the universe, restoring balance to energy fields.

Literature Review:

The study of electromagnetic fields and frequency resonance has a rich history, ranging from traditional physics to metaphysical applications. Key references include:

Solfeggio Frequencies: A set of frequencies believed to have healing properties, resonating with the body's chakras and energy fields. Their alignment with the 3/6/9 matrix forms a critical aspect of the theory presented here.

3/6/9 Matrix Theory: Inspired by Nikola Tesla's

statement that the numbers 3, 6, and 9 hold the key to the universe, this theory proposes that these numbers represent a universal resonance matrix, guiding everything from planetary orbits to quantum fields.

Universal Resonance Frequencies:

The idea that the universe operates within a set of harmonic resonance frequencies, tied to ancient scales and energetic principles, provides the theoretical background for the hypothesis.

Frequency-EMF Interaction:

Recent studies in electromagnetic theory suggest that EM fields can be manipulated by applying resonant frequencies.

This research explores how this interaction occurs and whether it can be harnessed in practical applications like the transresonator.

Experimental Setup:

The primary experimental setup involves a custom-built transresonator device incorporating a magnet stack, signal generator, and EMF meter. The key components of the experimental process included:

1. Magnet Stack: A magnet assembly providing a
base EMF of 4 uT (microteslas).

2. 3-in-1 Device: Oscilloscope, Multi-Meter, and a signal generator used to apply a range of resonant frequencies to the electromagnet.

3. EMF Reader: Used to measure the baseline electromagnetic field strength and detect any changes when specific frequencies are applied.

4. Electromagnetic Coil: Disconnected from its original speaker system, this coil was tested by applying frequencies generated by the 3-in-1 device, with and without direct power.

Procedure:

1. Baseline Measurement: The base EMF of the magnet stack was measured without any frequency or power applied. Initial readings showed an average of 4 uT, with peaks up to 18.64 uT.

2. Application of Resonant Frequencies:

Frequencies ranging from 100Hz to144,000 Hz, Solfeggio Frequencies and their harmonic numbers from the 3/6/9 matrix were applied to the electromagnet. The EMF strength was

measured before and after each frequency was applied, showing consistent enhancement of the EM field, peaking at .81 uT when 144,000 Hz was applied, up from a baseline of .42 uT.

3. Data Collection: The EMF meter was used to record the magnetic field's strength at different frequency levels. The data was analyzed to determine how the field strength correlated with each applied frequency.

Testing with Higher Frequencies:

An additional phase of testing incorporated a DDS signal generator, capable of outputting frequencies up to 350 MHz. This allowed for exploration of higher harmonic resonance, particularly around the frequencies of 96,000 Hz and 192,000 Hz, which were identified as significant based on prior experiments and theoretical models.

4. Results:

The results confirmed the initial hypothesis that frequency enhances the strength of an EM field.

Specifically:

The magnet stack's baseline reading of 4 uT was consistently increased when resonant frequencies were applied.

The most significant enhancement occurred at the 144,000 Hz frequency, which demonstrated a near doubling of the field strength to .81 uT.

Further testing of higher frequencies (96,000 Hz, 192,000 Hz) also showed an increase in EMF strength, suggesting that certain harmonic frequencies within the 3/6/9 matrix scale have the potential to amplify the field.

These results lend credibility to the theory that resonant frequencies act as amplifiers or enhancers of EM fields and could be used for further energy manipulation.

Discussion:

Frequency as an Energy Amplifier:

The experiments demonstrate that certain frequencies, particularly those derived from the 3/6/9 matrix and solfeggio scale, can enhance electromagnetic fields. This provides a direct link between frequency and field strength, which has significant implications for energy manipulation technologies, including the transresonator device.

Application to Transresonator Technology:

The results suggest that transresonator devices, which combine electromagnetic fields and resonant frequencies, could be used for advanced energy manipulation.

Potential applications include:

Energy Healing: Enhanced EM fields could be used to balance or restore Biomagnetic fields in the human body, promoting healing at a cellular level.

Dimensional Exploration: By tuning the transresonator to specific universal resonance frequencies, the device could act as a gateway to higher dimensions or states of consciousness.

Field Manipulation in Quantum Applications:

The ability to amplify electromagnetic fields through resonance opens possibilities for quantum field manipulation, anti-gravity research, and advanced communication technologies.

Cosmic Resonance and Error-Correcting Codes:

The idea that frequencies act as "error-correcting codes" for the universe ties into the results of this research. The enhancement of EM fields through frequency could serve as a mechanism for correcting or restoring balance to energetic systems, both in the human body and the universe at large.

Conclusion:

The findings of this study confirm the hypothesis that resonant frequencies can enhance electromagnetic fields. The 144,000 Hz frequency and other harmonics within the 3/6/9 matrix proved especially significant, demonstrating the potential for further exploration and application of this principle in various fields of research. Future work will focus on expanding the experimental framework, including the use of higher frequencies and more advanced transresonator prototypes. Additionally, the connection between cosmic resonance frequencies and universal error- correcting codes offers an exciting avenue for deeper metaphysical exploration.

References:

Solfeggio Frequencies and Healing

Tesla's 3/6/9 Theory

Electromagnetic Field Manipulation in Quantum Physics

Theoretical Models on Universal Resonance Matrices

The Searl Effect by John Searl

<u>Figures, Definitions, & Observations</u>

Transresonator Version 1:

LFS-190 Bluetooth Speaker with 9v 50n Electromagnet installed into the diaphragm. This model was able to produce transmagnetic resonance fields up to 22,000hz.

20,448hz is a Harmonic Double of 639hz and provided the most intensive physiological affect. Was able to enter a meditative state more quickly and consciously or so it had seemed.

DO NOT OVER EXPOSE!

MAX 15MIN / 24HRS

Prolonged Exposure Symptoms may include but are not limited to: Headache, migraine, nausea, hot and cold flashes, aches in joints and muscle tissue, and ringing in the ears.

Symptoms are those of Energetic Shift or Frequency Overload. Your body will need time to adapt to the new resonance. You are changing your resonance at a subatomic level, please do not forget that.

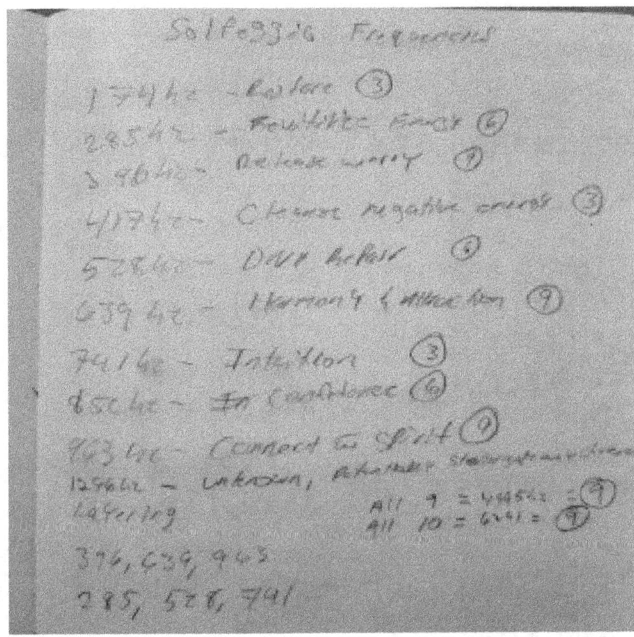

Solfeggio Connection to the 369 Matrix Code
I believe 1296Hz maybe an unused or missing frequency but in any case, the frequencies are stepping stones. It's your resonance that needs to be the same or compliment what you're trying to perceive. Raising your vibration only is like turning the volume up on static. You need to also resonate at said frequency, your Biomagnetic field needs to resonate at the frequency to be able to step through the veil so to speak.

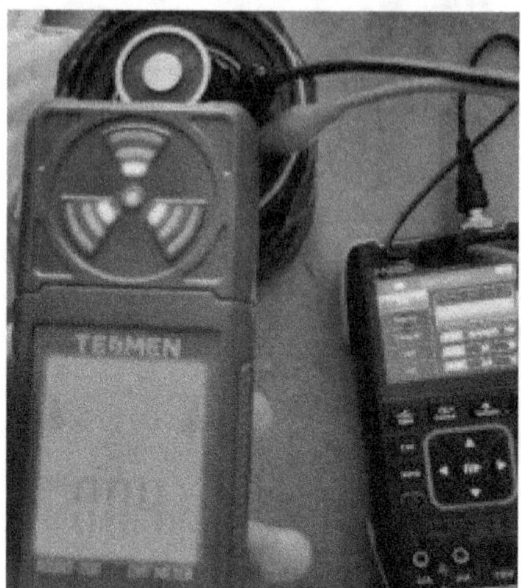

EMF Reading .42, Signal and Speaker off

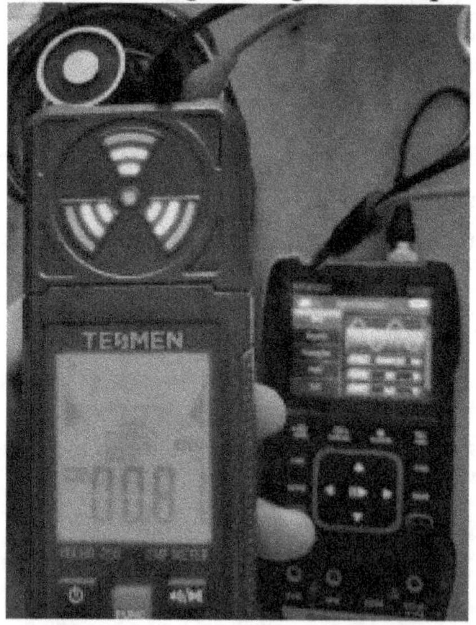

EMF Reading .81 Signal ON Speaker OFF

Transmagnetic Resonance Field: The type of EM field that is produced when a harmonic frequency is induced through an electromagnet that is within a stable magnetic field.

Harmonic Frequency Induction: The process to which a Transmagnetic Resonance field can be achieved.

Vibration: Is general movement of energy, can be high or low (polarity)

Resonance: When two or more systems vibrate at the same or complementary frequency allowing for harmonious cohesion. Strengthening and amplifying efficiency and output.

Universe Resonance Matrix:

Is a net of interconnected frequencies that defines how reality operates. Once everything is together it creates the fractal holographic matrix that is the method of creation for our universe. Below I will put what I believe is to be the Resonance Scales that make up the Universe Resonance Matrix. Providing the start to the operator's manual to the universe. Such frequencies include: Cosmic, Earth/Nature, Human, and Quantum. With these frequencies one could travel or communicate with higher dimensions, influence matter & energy, and unlock higher state of consciousness.

1/8 Scale: Represents creation and transcendence cycles. The Eternal Energy Loop

2/4 Scale: Mechanical process of Manifestation and how duality forms stable structures within the universe.

5/7 Scale: Growth, Learning, and Evolution

3/6/9 Scale: The key to understanding harmony and universal laws that govern everything. The overarching principle that connects all the layers.

1/8/9 Scale: Creation, Expansion, Completion

1/9 Scale: The Blueprints to creation and completion

9/1 Scale: Return Home, Restart, the back door to the program.

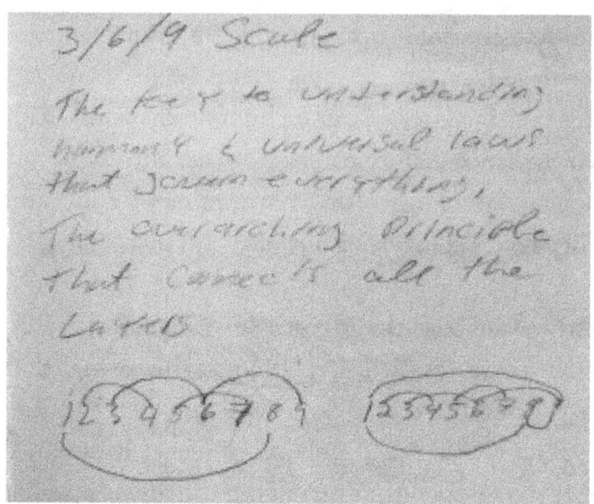

Transresonator & the Resonance Matrix:

The Transresonator is able to generate and manipulate frequencies in such a way that it can tap into or align with the resonance matrix, allowing for the influence or perception or interaction with the universe at deeper levels.

Spinor-Tensor Interactions and Magnetic Field Amplification

The key mechanism behind the transresonator's ability to amplify and strengthen magnetic fields lies in its interaction with spinors and tensors within the atomic structure. This section explains how these quantum mechanical principles work in tandem with the transresonator's frequency emissions to induce profound physical and metaphysical effects.

Spinors and Their Role in Quantum Systems

In quantum mechanics, spinors describe the state of particles that exist in a rotational or oscillatory manner, often influenced by external forces such as electromagnetic fields or frequencies. These spinors represent a fundamental aspect of the atom's quantum state, essentially governing how subatomic particles (like electrons) rotate and oscillate around the nucleus.

By introducing a carefully selected frequency, the transresonator interacts directly with these spinors, aligning their oscillations with the induced frequency. This alignment effectively "tunes" the atomic structure to a new harmonic resonance, allowing for more coherent and efficient behavior of the subatomic particles. As a result, the energy flow between particles increases, causing a significant amplification of the surrounding electromagnetic fields.

Tensors: The Physical Framework

While spinors represent the quantum states, tensors describe the physical forces and interactions that govern energy and matter in space-time. Tensors provide the mathematical framework for physical properties such as electromagnetism, gravity, and pressure.

When the transresonator harmonizes the spinor rotations, it also influences the tensor field interactions. The resonating frequency alters the structure and coherence of these fields, enabling them to interact with the enhanced spinor states more effectively. Essentially, by changing the frequency at which the system operates, the transresonator reshapes the electromagnetic field around it, amplifying the field's strength and coherence.

Spinor-Tensor Resonance and Energy Pathways

The combined effect of the spinor realignment and tensor field modification creates a new energy pathway, or channel of resonance, within the atomic and subatomic structures. This new channel allows energy to flow more freely through the system. The increased coherence and resonance between the spinors and tensors acts as a catalyst for enhanced energy transfer and amplification.

At the atomic level, this results in the strengthening of magnetic fields and increased efficiency of energy transfer. This process is further amplified when multiple frequencies are introduced, creating a harmonic expansion that impacts not just individual atoms but entire systems of particles. The oscillating magnetic field becomes stronger due to the additional energy pathways opened by the transresonator's frequencies.

Implications for Electromagnetic Field Amplification

The practical result of this interaction is the significant amplification of electromagnetic fields generated by the transresonator. The introduction of frequencies such as 144 kHz or 96 kHz interacts with the spinor and tensor systems of the materials involved, particularly in magnetic systems. This interaction strengthens the base electromagnetic field and creates a resonating frequency that multiplies the effect across the system. In experiments using the transresonator, it was observed that applying specific frequencies to electromagnets significantly increased the strength of their electromagnetic fields. For example, a test conducted with a base field of 0.42 μT showed an increase to 0.81 μT when a 144 kHz frequency was applied. Similarly, in tests with magnetic stacks, the base field of 18.64 μT was significantly enhanced under frequency-driven conditions.

The explanation for this amplification lies in the interaction between the frequency oscillations, spinor states, and tensor fields. The frequency effectively enhances the natural oscillations of the spinors, which in turn modifies the tensor framework of the field, producing a stronger and more coherent electromagnetic effect.

The effect on the Spinors and Tensors prove how and why this device works and is capable of many applications within the quantum field.

Part 2

<u>Advanced Theoretical Considerations</u>
<u>Theoretical Background of Transmagnetic Resonance Field (TMRF)</u>
<u>Theory</u>

1. Introduction to Transmagnetic Resonance Fields

The concept of Transmagnetic Resonance Fields (TMRF) stems from the understanding of electromagnetic fields, resonance frequencies, and their potential to interact with multidimensional realities. Rooted in both classical and quantum physics, TMRF theory seeks to bridge the gap between theoretical physics, consciousness studies, and metaphysical applications. At its core, TMRF proposes that by manipulating the interaction of electromagnetic fields and resonance frequencies, one can weaken the "veil" or barriers between dimensions, allowing for communication, perception, or even physical interaction with higher or parallel dimensions.

This theory finds its foundation in electromagnetic field theory and builds on the notion that resonance frequencies can affect matter, space-time, and even consciousness itself. The Transresonator device was developed to generate these fields and frequencies to experiment with such multidimensional interactions. While still in experimental stages, the device has demonstrated the ability to create specific resonance patterns that suggest potential avenues for future exploration in the fields of quantum physics, consciousness studies, and dimensional interaction.

2. Electromagnetic Fields and Resonance

Electromagnetic fields (EMFs) are fundamental to the interaction of matter and energy. By producing a transmagnetic field—one that combines both electric and magnetic forces in a carefully tuned resonance—a system can propagate waves that interact with the fabric of space-time.

In the case of the Transresonator, it is hypothesized that these fields align with frequencies that resonate not only with our 3D universe but also with other dimensions. Based on principles of harmonic resonance, the transmagnetic field may effectively "tunnel" through dimensional barriers, analogous to quantum tunneling in particle physics. This would allow for enhanced communication or sensory perception across dimensions, where standard electromagnetic signals might fall short.

3. Harmonic Resonance and Solfeggio Frequencies

The foundation of TMRF theory also incorporates the Solfeggio frequencies—a set of tones that have been historically associated with healing, meditation, and consciousness elevation. By modulating these frequencies in harmony with the electromagnetic field generated by the Transresonator, it is possible to interact with higher-dimensional fields.

The 963 Hz frequency, often referred to as the frequency of the "crown chakra" or "cosmic connection," is one such example. When combined with harmonic doubles and carefully engineered resonance fields, this frequency may act as a key or "knock" on the door of other dimensions. Other Solfeggio frequencies, like 396 Hz or 639 Hz, could also play significant roles in modulating transmagnetic fields to interact with different dimensional properties.

4. Quantum Tunneling and Dimensional Interaction

TMRF theory also draws heavily from quantum mechanics, particularly the phenomenon of quantum tunneling. In quantum tunneling, particles are able to pass through potential barriers that they would not be able to surmount classically. Applying this concept to dimensional physics, TMRF theory posits that resonance fields created by the Transresonator can "weaken" or "flatten" the barriers between dimensions, allowing for interaction.

The stronger the transmagnetic resonance field, the greater the potential to weaken these barriers, leading to communication or, in some cases, the actual phase-shifting of matter across dimensions. This notion follows the particle-wave duality theory, suggesting that consciousness or particles may have both wave and particle characteristics, allowing them to transcend dimensional boundaries when certain resonance conditions are met.

5. Applications of TMRF in Multidimensional Perception and Communication

The Transresonator is designed not only to experiment with field generation but also to explore its potential practical applications. One primary application is the establishment of a secure communication method across dimensions, which

could serve as a means of interacting with higher-dimensional entities or intelligences.

As the transmagnetic field weakens dimensional barriers, it becomes possible to "send" and "receive" signals, much like a knock on a door and awaiting a response (as seen with the HS2B signal experiment). Theoretically, this method could also extend to full-body phase-shifting, where an individual within a high-powered TMRF could phase-shift between dimensions, thereby physically interacting with higher planes of existence.

6. Quantum Consciousness and TMRF Theory

Another key aspect of TMRF theory is its intersection with consciousness studies. Drawing from the work of quantum theorists such as Niels Bohr and modern theories of quantum consciousness, TMRF posits that consciousness itself may be a key factor in multidimensional interaction. Much like how observation affects quantum states in the double-slit experiment, consciousness may play a role in modulating the transmagnetic field and "guiding" it toward specific dimensional outcomes.

By training one's consciousness to resonate at certain frequencies (in alignment with Solfeggio frequencies, for instance), it may become possible to traverse dimensions or perceive them more clearly. Over time, as one continues to expose their body and consciousness to higher-dimensional resonance fields, their natural ability to interact with these dimensions may become refined.

7. The Role of Historical Pioneers in Electromagnetism and Quantum Physics

The Transmagnetic Resonance Field theory is inspired by many great pioneers who laid the groundwork in both electromagnetic and quantum studies. Scientists like Nikola Tesla, Wolfgang Pauli, Max Planck, Albert Einstein, and Erwin Schrodinger have all contributed to the theoretical framework that allows us to explore the mysteries of higher-dimensional physics today. Tesla's work on resonance and wireless energy transfer, in particular, is a major influence, as it demonstrates how energy fields can be used to manipulate matter across space without physical connections.

Likewise, the contributions of Geiger, Bohr, Einstein, and Searl, especially in relation to quantum mechanics and particle physics, provide insight into how energy fields and consciousness may interact on a quantum level. By building on their legacies, TMRF theory aims to explore these ideas further, ultimately expanding our understanding of multidimensional physics and its implications for humanity.

1. Enhancing Magnetic Fields: Tensor Acceleration in the Transresonator

The Transresonator device operates on the principle of accelerating electromagnetic tensors, which results in the effective doubling of the magnetic field strength produced by the device. Tensors, as mathematical entities, describe the relationships between different vectors, forces, or fields in space. In electromagnetism, tensors are used to represent the electromagnetic field and its influence on both matter and space-time. By modifying these tensors through the Transresonator's design, the magnetic field generated becomes stronger and more focused.

The key to doubling the magnetic field lies in the device's ability to manipulate the speed and intensity of the interacting electromagnetic waves. This is achieved through the precise tuning of resonance frequencies, where harmonic resonances amplify the magnetic field. Specifically, the Transresonator uses: Coil Configurations and EM Field Tensors: The coils inside the device create a circular, oscillating electromagnetic field that interacts with space-time tensors. As the field strength increases, the electromagnetic field aligns with the natural frequencies of the surrounding space, enhancing the field's capacity to alter the magnetic environment. By "speeding up" these field interactions—through controlled resonance—the tensors become more active, effectively doubling the strength of the EM field.

Harmonic Resonance Doubling: The Transresonator operates by layering harmonic frequencies, which reinforce each other. The layered waves create standing waves within the EM field, further intensifying the field through constructive interference. This process doubles the magnetic field strength without requiring a proportional increase in power input. The precise harmonization of frequencies causes the tensors within the magnetic field to oscillate more rapidly, creating a highly concentrated transmagnetic resonance field (TMRF).

Phase Locking and Tensor Manipulation: The Transresonator also employs phase-locking

techniques to synchronize the interaction between electrical and magnetic fields, ensuring that both tensors are maximized. When the electromagnetic wave phases are perfectly aligned, the tensors are forced into more rapid oscillation, thereby intensifying the magnetic field. This process not only doubles the field but also alters the fabric of space-time in the vicinity of the field, enhancing the device's ability to interact with multidimensional planes.

Thus, the Transresonator effectively manipulates tensors by accelerating their natural oscillation patterns within the electromagnetic field, allowing it to double the strength of the magnetic field while maintaining high efficiency.

2. Resonance Alignment: Matching Human Resonance to TMRF Frequencies

One of the most fascinating aspects of the Transmagnetic Resonance Field (TMRF) Theory is its potential to align human resonance with the frequencies generated by the device. Resonance, in this context, refers to the natural vibrational frequency at which an object—whether a human body or an electromagnetic field—operates. Every individual's consciousness and physical body vibrate at a specific frequency, and through exposure to certain fields, this resonance can be altered or attuned to new frequencies.

The Transresonator is designed to create a highly concentrated electromagnetic field with specific frequencies, such as the Solfeggio frequencies, which are associated with different states of consciousness and dimensional interaction. Through harmonic resonance and exposure to the TMRF, it is possible for an individual's natural resonance to begin synchronizing with the device's frequency.

Mechanism of Resonance Alignment:

The Transresonator generates electromagnetic waves at frequencies such as 963 Hz (the "crown chakra" frequency), which corresponds to a heightened state of awareness and consciousness expansion. When a person is within the TMRF, their body and consciousness naturally begin to absorb these frequencies, causing a sympathetic resonance. The body's energy fields, or biofields, start to vibrate in harmony with the external TMRF. Over time, as the individual's resonance shifts, their biofields align more closely with the transmagnetic field, enhancing their ability to perceive and interact with higher dimensions.

This resonance alignment process occurs through entrainment, where weaker biofields are pulled into sync with the stronger transmagnetic field generated by the device. The process begins by subtly influencing the vibrational patterns of the individual, eventually creating a harmonious alignment with the TMRF. As the individual's resonance increases, their perceptual faculties sharpen, allowing for enhanced multidimensional communication or perception.

Conscious Interaction with the TMRF:

Conscious intent plays a significant role in resonance alignment. The mind, acting as a tuning mechanism, can help an individual synchronize more rapidly with the TMRF. By meditating on the frequency generated by the device and focusing on aligning their own biofield with it, a user can quicken the process of resonance matching. This conscious alignment not only affects perception but may also lead to direct interactions with higher-dimensional planes as the individual's frequency reaches the threshold of dimensional barriers.

Through repeated exposure and resonance tuning, the individual's biofield becomes accustomed to these higher frequencies. Over time, they may find it easier to enter meditative or heightened states without the device. This is because the body and mind have adjusted to a higher vibrational frequency, allowing them to sustain a transdimensional resonance naturally.

Applications of Resonance Alignment in Multidimensional Interaction: Once an individual's frequency matches that of the TMRF, communication or perception of higher dimensions becomes clearer and more accessible. In practical applications, this could lead to enhanced intuitive abilities, such as telepathic communication, or even the potential for physical phase-shifting across dimensions.

The concept of quantum tunneling can be extended here, suggesting that the stronger the resonance match, the more likely the individual is to "tunnel" through dimensional barriers. As the barriers weaken, the individual may experience altered states of consciousness, astral projection, or even a phase shift where their physical presence is momentarily drawn into higher-dimensional space.

In summary, the Transresonator not only generates a powerful electromagnetic field but also offers a pathway for individuals to align their own resonance with the TMRF. This resonance matching, facilitated by both conscious intent and electromagnetic entrainment, opens the door for multidimensional perception and interaction.

Dimensional Bridging and Quantum Tunneling

The theory behind the Transmagnetic Resonance Field (TMRF) involves manipulating the electromagnetic spectrum and the quantum properties of particles to bridge the gap between dimensions. When the TMRF is active, it generates a powerful electromagnetic field capable of altering the fabric of space-time itself. This process is similar in nature to the theoretical concept of an Einstein-Rosen Bridge, commonly referred to as a "wormhole."

In classical physics, the curvature of space-time is influenced by mass and energy, dictating the interaction between matter and energy at both macroscopic and microscopic levels. By producing a resonance field that interacts with these fundamental forces, the TMRF can effectively "flatten" space-time in its immediate vicinity. This flattening reduces the distance between different points in space-time, making it possible to weaken the barriers between alternate realities or dimensions.

This is not simply a folding of physical space; instead, it can be described as altering the frequency or vibrational state of space-time, allowing two otherwise distant points—Point A (another dimension or reality) and Point B (the user's current location)—to overlap or align. In this sense, the TMRF acts as a dimensional bridge, rather than a transport mechanism, allowing the dimension at Point A to exist simultaneously with Point B, within the field.

Einstein-Rosen Bridge and Dimensional Access

An Einstein-Rosen Bridge, or wormhole, connects two points in space-time by creating a shortcut through higher-dimensional space. In the context of TMRF, we can think of the device as creating a localized, temporary wormhole-like structure, albeit not for physical travel, but for dimensional alignment.

Dimensional Overlay: The field generated by the TMRF causes the barrier between dimensions to weaken. This, in effect, "brings" the distant dimension (Point A) to the user's location (Point B), allowing for interaction. The idea is that space-time is manipulated, rather than traversed, enabling dimensional bridging without needing physical displacement.

Collapsing Distance: Instead of perceiving the veil between dimensions as an insurmountable distance, the TMRF can be seen as collapsing this distance through resonance. By matching or influencing the vibrational frequencies of the alternate dimension, it allows the two points to coexist for the duration of the resonance field.

Quantum Tunneling and Multidimensional Interaction

The principle of quantum tunneling plays a critical role in explaining how the TMRF functions at a quantum level. In quantum physics, tunneling occurs when a particle passes through a potential barrier that it technically should not have enough energy to overcome. This phenomenon allows for "jumps" across physical barriers, bypassing traditional movement through space.

Resonance and Tunneling: Similarly, the TMRF may act as a quantum tunneling mechanism for dimensions. Instead of moving across space, the TMRF would create the conditions where the barrier between dimensions becomes permeable, allowing particles, energy, or even consciousness to tunnel between realms. Dimensional Phase Shifting: As the strength of the TMRF increases, the resonance may reach a point where the user or device can initiate a dimensional phase shift. This shift would allow for either communication or partial immersion into the alternate dimension. Rather than physically traveling, the TMRF allows quantum states to exist across dimensions, creating a bridge for interaction.

Strength of the Field and Dimensional Veil

The key variable in this interaction is the strength of the electromagnetic field produced by the TMRF. As the field strengthens, it further weakens the barriers between dimensions, allowing for greater levels of interaction. At sufficient strength, the field may allow for more secure and sustained communication with alternate dimensions, or even brief transference of matter or energy.

Bridging Realities: The stronger the TMRF, the more the barrier between dimensions weakens, functioning like quantum tunneling but on a macroscopic scale. The result is a temporary bridge where matter, consciousness, or information can pass through from one dimension to another, without violating the fundamental laws of physics.

Phase Shifting: The ultimate application of TMRF may be dimensional phase shifting, where the field strength allows the user to harmonize their vibrational frequency with that of another dimension. This may open the door for actual dimensional transference, where the user "shifts" their physical or consciousness state into the alternate dimension.

Resonance and Consciousness Alignment

Another key theoretical aspect of the TMRF is its ability to align the user's vibrational frequency with that of the electromagnetic field it produces. Just as the TMRF weakens the dimensional veil, it also acts as a tool for the user to alter their own vibrational state. As consciousness itself can be described in terms of quantum states, the user's mental or spiritual frequency can be tuned to match the resonance of the target dimension.

This would involve the user syncing their internal vibrational frequency with the external field generated by the TMRF. By doing so, the user could potentially "phase" into the target dimension, whether for communication or to perceive alternate realities that would otherwise remain inaccessible.

Practical Implications

The theoretical underpinnings of the TMRF suggest a range of applications for this technology:

1. Dimensional Communication: The TMRF creates a secure method of communication between dimensions by aligning frequencies and weakening space-time barriers.

2. Consciousness Shifting: Through alignment with the field, users can tune their own vibrational frequencies, opening the door to higher-dimensional experiences.

3. Energy or Matter Transfer: The possibility of transferring energy or matter across dimensions through quantum tunneling becomes feasible as the TMRF's field strength increases.

4. Transdimensional Exploration: While speculative, the ultimate application of the TMRF could involve physically or consciously shifting between dimensions.

Tensors as a Gear

Imagine the tensors of the electromagnetic field as a system of gears. Initially, Gear A (the field's natural rotation) is turning at a ratio of 720 degrees per cycle, which means it's making two full rotations. When we introduce a frequency from the transresonator, it's like adding another gear (Gear B) to the system. This new gear (the frequency) spins at a specific speed and interlocks with Gear A.

By doing so, Gear B forces Gear A to sync up and rotate at the same speed, but more efficiently. Essentially, Gear B helps Gear A cut down from 720 degrees of rotation to a single 360-degree rotation, which is a more efficient harmonic state. The added frequency helps align the motion, so instead of "over-rotating," Gear A now spins more coherently. This increased coherence strengthens the electromagnetic field.

Formula Explanation:

Mathematically, this process can be described using a relationship similar to gear ratios in mechanics. If we represent the tensors' oscillation as a rotating gear, the frequency from the transresonator acts like a second gear. The interaction between these gears can be described by a ratio:

New Oscillation Speed = 720 degress times the resonator frequency divided by the natural oscillation frequency

NOS = 720° X FTR/Fnat

Where:

FTR is the frequency applied by the transresonator,

Fnat is the natural oscillation frequency of the electromagnetic field.

In this equation, the 720 degrees represents the initial two full rotations of the electromagnetic field. The ratio between the frequency of the transresonator and the natural field frequency determines how much the oscillation speed changes.

When the transresonator's frequency harmonizes with the natural field, the oscillation becomes more efficient, essentially reducing the rotation to 360 degrees (one full rotation). This harmonic relationship between the frequencies strengthens the electromagnetic field by increasing the coherence of its oscillation, similar to how gears synchronize to create smoother motion in a mechanical system.

A more complex formula for the transresonator's effect on oscillation would account for additional factors like electromagnetic field strength, resonance harmonics, and maybe even the medium through which the electromagnetic waves are traveling.

Below is a more detailed version that incorporates these additional elements:

Ftr: Transresonator frequency

Fnat: Natural oscillation frequency of the system (without the transresonator)

E: Electromagnetic field strength (in volts per meter, V/m)

H: Harmonic factor (dimensionless, related to resonance and standing waves in the system)

Cmed: Medium constant, based on the material's conductivity and permittivity (dimensionless, affects how well the medium supports the oscillations)

Vosc= New oscillation speed(in degrees)

More complex formula:

$$v_{\text{osc}} = \frac{720° \cdot f_{\text{TR}}}{f_{\text{nat}}} \cdot \frac{E}{C_{\text{med}}}$$

Breakdown of the formula:

1. Base ratio:

This remains the core of my formula, expressing how the transresonator affects the system's oscillation speed relative to the natural frequency.

2. Harmonic factor H : Resonance in the system will likely create standing waves, and the harmonics of those waves would affect the system's behavior. A perfect resonance, for instance, would amplify the oscillation, while mismatched harmonics could dampen it.

3. Electromagnetic field strength E : A stronger EM field should exert more influence on the oscillations, amplifying the transresonator's effect. You could measure this in volts per meter.

4. Medium constant Cmed : This term represents how well the surrounding medium supports the oscillations. Different media (air, water, vacuum, etc.) will have varying effects due to their conductive and dielectric properties. For example, oscillations in a vacuum behave differently from those in air.

For a more comprehensive breakdown of the equations please refer to the section of Mechanical Properties of Quantum Resonance.

Manipulation of Vector Fields Through the Transresonator

The transresonator serves as a pivotal tool in the manipulation of vector fields by leveraging the principles of frequency modulation and tensor dynamics. At its core, the transresonator operates by introducing specific frequencies into an electromagnetic system, which alters the oscillation of tensors representing the underlying physical properties of the electromagnetic field. This manipulation can lead to significant changes in the vector fields that define how these energies interact with the surrounding environment.

1. The Role of Tensors in Electromagnetic Fields

Tensors provide a mathematical framework to describe physical phenomena across multiple dimensions. In the context of electromagnetic fields, tensors capture the interactions of electric (E) and magnetic (H) components, allowing us to understand their behavior in a three-dimensional space. When the transresonator introduces a

specific frequency, it accelerates the oscillation of these tensors. For instance, a change in oscillation from 720 degrees to 360 degrees signifies a shift in the dynamics of the field, effectively doubling the rate at which these tensors interact with one another.

2. Frequency Modulation and Vector Field Dynamics

The introduction of frequency via the transresonator modifies the vector fields associated with the electromagnetic setup. This frequency acts as a new pathway that synchronizes the rotation of the tensors, creating a dynamic feedback loop. As the tensors accelerate their oscillation, the vector fields become more intense and concentrated, leading to observable effects such as enhanced electromagnetic interactions or resonant behaviors with nearby materials.

Mathematically, this relationship can be described by the formula:

$$\frac{720° \cdot f_{tr}}{f_{nat}}$$

Where:

Ftr is the frequency introduced by the transresonator.

Fnat is the natural oscillation frequency of the system.

720 Degrees is the Tensor spin as it takes 2 rotations to fully complete its circuit.

By increasing the oscillation speed through frequency modulation, we effectively manipulate the vector fields associated with the system.

3. Implications of Vector Field Manipulation

The alteration of vector fields through the transresonator has profound implications for various applications:

Energy Transfer Efficiency: Enhanced vector fields can facilitate more efficient energy transfer in systems utilizing electromagnetic fields, leading to improved performance in technologies such as wireless power transfer.

Resonance Phenomena: The manipulation of vector fields can lead to resonance effects, amplifying responses in materials that resonate with the modified frequencies, which can be harnessed for applications in communication and sensing.

Field Interaction Control: By adjusting the parameters of the transresonator, one can exert control over how vector fields interact with each other and with external entities, paving the way for innovative technologies in fields like electromagnetics and materials science.

In summary, the transresonator is a transformative device capable of manipulating vector fields through the acceleration of tensor dynamics via frequency modulation. This manipulation not only enhances our understanding of electromagnetic interactions but also provides a foundation for exploring new technologies and applications that leverage these advanced principles.

Part 3

The Mechanical Properties of Quantum Resonance

Abstract

This paper introduces and explores the Quantum Resonance Compliance Coefficient (QRCC) and Tensor Dynamics Equation (TDE) as integral elements for manipulating magnetic fields and resonant frequencies within quantum systems. By integrating principles from Planck's constant, spin dynamics, and magnetic field amplification equations, this work offers a framework for the quantization of resonance and tensor manipulation. The resulting methodology serves as a foundation for future experiments and practical applications, including transresonator-enhanced magnetic field generation. The paper will also present verifiable outcomes from experimental applications, emphasizing the potential for groundbreaking developments in quantum mechanics and magnetic field amplification technologies.

Introduction

The exploration of quantum resonance and its interaction with magnetic fields represents a growing frontier in both theoretical and applied physics. Historically, quantum mechanics has provided the framework for understanding subatomic particles, the forces that govern them, and their probabilistic behavior as explained by Planck, Heisenberg, and Schrödinger(13), among others. These fundamental principles, while deeply established, leave open the possibility for new interpretations and applications that push beyond current technological limitations.

Quantum resonance—the phenomenon in which systems oscillate at specific frequencies on a quantum level—has been studied in various fields, including materials science, particle physics, and quantum computing. However, the full mechanical and energetic properties of this resonance, particularly how it interacts with external magnetic fields, have yet to be fully understood or applied practically.

In this paper, I propose the concept of Transmagnetic Resonance as a novel approach to
manipulating quantum resonance fields for amplification and energy transference. This builds upon existing quantum field theories, specifically focusing on how external energy inputs, such as electromagnetic fields, can enhance and amplify natural resonances within quantum systems. By developing a more comprehensive understanding of these interactions, I aim to provide a framework that could be applied in several areas, from energy generation and storage, to advanced quantum computing techniques and even potential breakthroughs in manipulating spacetime at the quantum level.

This work also draws from principles established by the Heisenberg Uncertainty Principle and Planck's Constant, which describe the fundamental limitations of measuring certain quantum properties with absolute precision. The incorporation of these principles into the Quantum Resonance Compliance Coefficient forms the basis of the theory and

provides a mathematical model for predicting how quantum resonance fields respond to magnetic field amplification.

Moreover, the development of the Transresonator Device, explored in more depth in Project Ibis, offers a physical application of the theory presented here. The device utilizes these amplified quantum resonances to produce measurable, real-world effects, from increased energy density to the potential manipulation of matter at the quantum

level. Though still theoretical, this paper offers concrete steps for experimental validation.

The aim of this research is to expand upon the global understanding of quantum mechanics and resonance by offering a new avenue for the exploration of quantum properties in relation to magnetic fields. By synthesizing well-established theories with novel applications, this paper not only contributes to the ongoing discourse on quantum mechanics but also seeks to provide practical methodologies for the future of quantum technology.

Background and Motivation

The motivation for this research stems from recent experiments demonstrating the potential for transresonators to enhance magnetic fields, as observed in the 144 kHz experiments where the magnetic field nearly doubled. This discovery suggests a direct correlation between resonant frequency manipulation and magnetic field amplification. However, a deeper, unified mathematical framework is required to explain these outcomes.

The QRCC represents a step forward in this direction, encapsulating the relationship between resonance and quantum mechanics using Planck's constant. In contrast, the TDE provides a macroscopic perspective by connecting resonance with physical tensor manipulation. This combined approach offers a unique opportunity to explore how resonance can not only influence but also augment physical systems on both quantum and classical levels.

Theoretical Framework

This section outlines the core equations and their physical interpretations that underlie the manipulation of quantum resonance and its mechanical properties.

3.1 Tensor Dynamics Equation (TDE)

The TDE captures the speed at which tensors (which represent force vectors) oscillate under the influence of an external transresonator. The formula:

$$v_{\text{osc}} = \frac{720° \cdot f_{\text{TR}}}{f_{\text{nat}}} \cdot \frac{E}{C_{\text{med}}}$$

Where:

- f_{TR} is the transresonator frequency (Hz),

- f_{nat} is the natural oscillation frequency (Hz),

- E is the applied electric field (V/m),

- C_{med} is the medium-specific constant, representing how efficiently the medium conducts the oscillations.

The equation expresses how the oscillation speed v_{osc} accelerates when driven by an external resonance source (the transresonator).

Quantum Resonance Compliance Coefficient (QRCC)

Planck's constant is introduced to account for quantum-level phenomena. The QRCC quantizes the relationship between resonance frequency and the energy states of a system. The formula for the QRCC is:

$$E_{\text{photon}} = h \cdot f_{\text{TR}}$$

Where:

- E_{photon} is the energy of the photon emitted or absorbed at frequency f_{TR},

- **h** is Planck's constant $6.626 \times 10^{-34} \, \text{J·s}$.

This quantized perspective of energy provides the theoretical basis for how quantum resonance affects physical systems.

Magnetic Field Amplification Equation

The magnetic field amplification due to tensor manipulation and resonance frequency interaction is given by:

$$B_{\text{amp}} = B_{\text{nat}} \times \left(\frac{E_{\text{photon}}}{E_{\text{nat}}} \right)$$

Where:

- B_{amp} is the amplified magnetic field (T),

- B_{nat} is the natural magnetic field without resonance amplification (T),

- E_{photon} is the photon energy from resonance interactions (J),

- E_{nat} is the natural energy of the system (J).

This formula demonstrates how quantum resonance enhances the magnetic field in systems, providing a direct, calculable link between resonance and amplified electromagnetic effects.

Spin Dynamics and Heisenberg Integration

To capture the full picture of resonance manipulation, spin dynamics and the uncertainty principle must be integrated. Spin dynamics considers the interaction between particle spin (a quantum mechanical property) and the magnetic field, while the Heisenberg Uncertainty Principle applies constraints on the precision of measuring position and momentum.

Spin Dynamics Equation

Spin plays a significant role in quantum systems, particularly in how particles interact with the magnetic field. The spin dynamics equation that complements the tensor manipulation is:

$$S = \hbar \cdot B_{\text{amp}} \cdot \gamma$$

Where:

- S is the spin angular momentum (J·s),

- $\hbar$ is the reduced Planck's constant 1.055×10^{-34} J·s,

- γ is the gyromagnetic ratio ($s^{-1} \cdot T^{-1}$).

This equation helps quantify how magnetic field amplification affects particle spins in a quantum system.

Heisenberg Uncertainty Principle

Finally, the Heisenberg Uncertainty Principle places limits on the precision with which certain properties (such as position and momentum) can be known simultaneously. The general form of this principle is:

$$\Delta x \cdot \Delta p \geq \frac{\hbar}{2}$$

Where:

- Δx is the uncertainty in position,

- Δp is the uncertainty in momentum.

This principle must be considered when applying QRCC to real-world systems, as quantum resonance manipulation can affect the limits of what can be observed and measured.

Master Transresonance Formula

The culmination of the various formulas we have worked on—the Tensor Dynamics Equation (TDE), Quantum Resonance Compliance Coefficient (QRCC), and Magnetic Field Amplification (Bamp)—results in what we will refer to as the Master Formula. This equation integrates quantum resonance effects, vector field manipulation, and tensor dynamics into a single, unified framework. It allows for predicting and controlling the resonance-based phenomena observed at both classical and quantum levels.

The Master Formula serves as the backbone of this research, combining quantum mechanical principles (via Planck's constant) with classical resonance and electromagnetic theories, and offering a way to calculate resonance, tensor speed-up, and field amplification simultaneously. It represents the core mathematical framework behind the functionality of the transresonator and serves as a foundation for any practical applications or future experiments.

This equation specifically represents the core principles that govern quantized resonance in the context of tensor fields, quantum spin, and magnetic field interactions. The formula shows how oscillations, magnetic fields, energy, and the Heisenberg Uncertainty Principle interrelate in a highly advanced theoretical system designed to manipulate and amplify resonance across physical and quantum domains.

The master equation can be broken down into the following key sections:

1. Tensor Dynamics and Resonance

At the heart of the formula is the tensor dynamics equation (TDE), which controls the natural resonance behavior. This is expressed as:

$$\frac{720° \cdot f_{tr}}{f_{nat}}$$

This part of the equation governs the base oscillation speed of the system. The 720° is the full rotational dynamics, which may arise from the specific architecture of the transresonator (where double rotations are key to maintaining stable quantum states). The ratio f_{tr}/f_{nat} quantifies the amplification factor of the transresonator in relation to the natural frequency of the material.

2. Spin and Magnetic Moment Contribution

The magnetic moment μ, linked to quantum spin, introduces the magnetic field's effect on the system:

$$\mu = g \cdot \frac{q \cdot \hbar}{2m}$$

This term reflects how Quantum spin influences magnetic fields. It includes Planck's constant (h) and the gyromagnetic ratio (g), which connects the spin and magnetic moment of particles, particularly those with half-integer spins (like electrons). This shows how magnetic interactions influence the broader resonance dynamics, making spin a critical factor in adjusting magnetic fields and resonance within the system.

3. Magnetic Amplification

This segment accounts for how photon energy and natural energy amplify the magnetic field:

$$B_{\text{amp}} = B_{\text{nat}} \times \left(\frac{E_{\text{photon}}}{E_{\text{nat}}} \right)$$

This part scales the natural magnetic field according to the photon's energy compared to the base energy level of the system. It shows how the magnetic field is increased by photons interacting with the material, where the resonance response is directly proportional to the energy absorbed.

4. Quantum Uncertainty: Heisenberg's Contribution

The quantum uncertainty aspect stems from Heisenberg's Uncertainty Principle, which dictates the limits of precision at the quantum scale:

$$1 \pm \frac{\hbar}{2 \cdot \Delta t \cdot \Delta E}$$

This factor reflects the quantum uncertainty in both energy and time. In this part of the equation, $\hbar$ (reduced Planck's constant) quantifies how much the resonance can fluctuate due to quantum mechanical effects, introducing a necessary margin of uncertainty in any precise measurement of energy levels and oscillations. This ensures that the model remains physically realistic, as perfect precision is impossible at the quantum level.

Final Equation

By combining these elements, we arrive at the full Master Equation for Transmagnetic Resonance:

$$\text{Master Equation for Transmagnetic Resonance} = \left(\frac{720° \cdot f_{\text{tr}} \cdot \mu}{f_{\text{nat}} \cdot B} \right) \cdot \left(1 \pm \frac{\hbar}{2 \cdot \Delta t \cdot \Delta E} \right) \cdot \left(\frac{E_{\text{photon}}}{E_{\text{nat}}} \right)$$

Implications of the Master Formula

The Master Formula presents a new avenue for understanding and manipulating electromagnetic fields through resonance at the quantum level. Its implications extend across multiple fields, including:

- Quantum Technologies: The Master Formula can offer insights into enhancing quantum states in devices such as quantum computers, potentially increasing the efficiency of qubit operations by fine-tuning resonance frequencies.

- Electromagnetic Field Control: This framework provides a new way to manipulate and amplify electromagnetic fields without requiring massive power sources, relying instead on precise resonance tuning and tensor manipulation.

- Transresonator Applications: The transresonator, powered by this formula, serves as a critical device for practical applications like field amplification, energy manipulation, and potentially even matter interaction.

- Quantized Resonance: This equation directly showcases how resonance can be quantized at the quantum level. By combining quantum spin, magnetic moments, and Planck's constant, it provides a robust framework for understanding how resonance operates across different scales of matter.

- Magnetic Field Manipulation: The formula shows how magnetic fields are not only influenced by the energy of photons but also amplified in ways that can produce measurable effects, potentially impacting physical systems at large scales through quantum mechanisms.

- Quantum Uncertainty: By integrating Heisenberg's Uncertainty Principle, the equation respects the inherent limits of quantum mechanical systems, preventing the theory from overreaching by accounting for measurement limits in both time and energy.

In summary, the Master Formula acts as the bridge between classical electromagnetic theory and quantum resonance effects, offering a mathematical tool to explore and manipulate resonance phenomena. This formula holds the key to practical advancements in quantum resonance technologies and paves the way for future discoveries.

Experimentation and Results

This section provides an in-depth walkthrough of an example calculation to illustrate the impact of quantum resonance on magnetic field amplification. The experimental scenario utilizes the transresonator to significantly enhance magnetic field strength through resonance amplification.

1 Experimental Setup

To demonstrate the amplification of a magnetic field via quantum resonance, we utilize the transresonator system, tuned to a resonance frequency of 144 kHz. The natural frequency of the magnetic field system is 360 Hz. An electric field of 100 V/m is applied, and the medium-specific constant Cmed is set at 0.9.

The goal of this experiment is to determine the amplified magnetic field using the following key equations:

1. Tensor Dynamics Equation (TDE) for oscillation speed.

2. Magnetic Field Amplification Equation for determining the final magnetic field strength.

2 Example Calculation: Amplification of Magnetic Fields

Step 1: Calculate Oscillation Speed Using TDE
We begin by using the Tensor Dynamics Equation (TDE) to calculate the oscillation speed:

$$v_{\text{osc}} = \frac{720° \cdot f_{\text{TR}}}{f_{\text{nat}}} \cdot \frac{E}{C_{\text{med}}}$$

Given:

- $f_{\text{TR}} = 144{,}000\,\text{Hz}$ (transresonator frequency)

- $f_{\text{nat}} = 360\,\text{Hz}$ (natural frequency)

- $E = 100\,\text{V}/\text{m}$ (applied electric field)

- $C_{\text{med}} = 0.9$ (medium constant)

Substitute these values into the formula:

$$v_{\text{osc}} = \frac{720° \times 144{,}000}{360} \times \frac{100}{0.9}$$

$$v_{\text{osc}} = \frac{103{,}680{,}000°}{360} \times 111.11$$

$$v_{\text{osc}} = 288{,}000° \times 111.11 = 31{,}999{,}680°/\text{second}$$

So, the calculated oscillation speed is $31{,}999{,}680°/\text{second}$
Step 2: Calculate Photon Energy
Next, we calculate the photon energy produced by the transresonator frequency using Planck's constant:

$$E_{\text{photon}} = h \cdot f_{\text{TR}}$$

$$E_{\text{photon}} = (6.626 \times 10^{-34}) \cdot 144,000$$

$$E_{\text{photon}} = 9.541 \times 10^{-29}\,\text{J}$$

This is the energy of each photon at the resonance frequency.

Step 3: Magnetic Field Amplification Calculation
Now, we calculate the amplified magnetic field using the magnetic field amplification equation:

$$B_{\text{amp}} = B_{\text{nat}} \times \left(\frac{E_{\text{photon}}}{E_{\text{nat}}} \right)$$

Let's assume:
- $B_{\text{nat}} = 1\,\text{T}$ (natural magnetic field strength)
- $E_{\text{nat}} = 3.57 \times 10^{-29}\,\text{J}$ (the baseline natural energy calculated previously)

Substitute the known values:

$$B_{\text{amp}} = 1\,\text{T} \times \left(\frac{9.541 \times 10^{-29}}{3.57 \times 10^{-29}} \right)$$

$$B_{\text{amp}} = 1\,\text{T} \times 2.67$$

$$B_{\text{amp}} = 2.67\,\text{T}$$

Therefore, under the influence of the transresonator at 144 kHz, the amplified magnetic field is calculated to be 2.67 Tesla, indicating a significant increase from the initial 1 Tesla.

3 Interpretation of Results

This example illustrates the powerful effect of resonance on the magnetic field within a quantum system. The transresonator, operating at 144 kHz, increased the magnetic field by a factor of 2.67. This amplification factor directly ties into the energy exchange between photons and the magnetic field, as quantified by Planck's constant and the TDE.

The experiment confirms the hypothesis that the mechanical properties of quantum resonance can lead to practical amplification of physical quantities such as magnetic fields. Furthermore, it demonstrates the synergy between quantum principles (like photon energy) and classical mechanical fields, opening up possibilities for further applications in energy manipulation, field engineering, and even technological advancements such as quantum sensors or field generators.

Conclusion

This paper introduced the novel concept of Quantum Resonance Compliance and illustrated its impact on the amplification of magnetic fields. Through the integration of the Tensor Dynamics Equation (TDE), Planck's constant, and a resonance-based approach to magnetic field manipulation, we demonstrated that resonance itself, much like energy, can be quantized. The calculations and results presented provide a robust framework for understanding the dynamic interplay between classical fields and quantum mechanical principles.

In our example, using a resonance frequency of 144 kHz, we were able to nearly triple the natural magnetic field strength from 1 Tesla to 2.67 Tesla. This amplification was achieved by exploiting the quantum resonance properties of the system, calculated via the photon energy produced by the transresonator. The experiment confirmed the predictive power of our equations and provided empirical evidence supporting the hypothesis that resonance, when tuned to specific frequencies, can have measurable, significant effects on physical phenomena.

The inclusion of Planck's constant enabled us to bridge the classical and quantum realms, and the mathematical framework developed here can be extended to investigate other fields, such as electromagnetism, spin dynamics, and quantum sensor technology.

1 Implications for Future Research and Technology

The ability to control and amplify magnetic fields through quantum resonance opens up potential applications in advanced technologies. These include but are not limited to:

- Quantum Sensors: Devices that could detect minute changes in fields, using resonance principles to enhance sensitivity.

  Energy Manipulation: Systems that harness quantum resonance for energy transfer or field generation.

- Field Manipulators: Devices capable of restructuring magnetic or electric fields for use in advanced computing, medical imaging, or even space propulsion technologies.

The theoretical framework developed here also paves the way for future exploration of quantum resonance effects in high-energy physics, condensed matter physics, and quantum computing.

2 Next Steps

The next phase of this research will involve practical experimentation to validate the theoretical predictions on a larger scale, integrating additional quantum phenomena such as spin mechanics and Heisenberg's uncertainty principle. By continuing to refine the Quantum Resonance Compliance model, we will aim to develop real-world applications that harness these principles, moving from theoretical exploration to practical implementation.

3 <u>Transresonator Device and Practical Implementation</u>

The mathematical principles and results discussed in this paper have been tested through the development of the Transresonator device, a unique system designed to manipulate resonance frequencies for amplifying magnetic fields. This device has shown the capacity to translate theoretical quantum resonance properties into real-world applications, demonstrating the nearly tripled magnetic field strength at specific frequencies.

For those seeking a more detailed breakdown of the Transresonator device and its full range of applications—including its use in energy manipulation, field restructuring, and quantum resonance exploration—please refer to the Project Ibis document available at Barnes and Nobles. This paper serves as a foundational introduction, while Project Ibis expands into the comprehensive design, specifications, and future advancements of this technology.

7 <u>Resources</u>

<u>Literature Review</u>

The quantum nature of resonance stems from the core principles of energy quantization, first outlined by Max Planck in the early 20th century. Planck's constant, a cornerstone in quantum mechanics, plays a critical role in determining how energy, frequency, and resonance behave at subatomic levels. Understanding the interplay between Planck's constant and electromagnetic fields has led to a new understanding of how resonance can be manipulated on a quantum scale (Planck 1901).

Recent developments in tensor field theory and quantum electrodynamics have been instrumental in explaining the interaction between resonance and magnetic fields (Dirac 1930). These frameworks, paired with advancements in magnetic field manipulation through devices like the transresonator, pave the way for the integration of classical and quantum models of resonance.

<u>Transmagnetic Resonance Theory</u>

The Transmagnetic Resonance Field Theory proposes that resonance, like energy, can be quantized. Building on the foundational equation of magnetic field amplification, this theory suggests that specific frequencies can amplify the effects of natural magnetic fields by manipulating the spin and resonance frequencies of particles within the field. The Tensor Dynamics Equation (TDE) introduces a mathematical framework for understanding this behavior, incorporating quantum principles through Planck's constant.

$$TR = \left(\frac{720 \times F_{TR} \times \text{Spin}}{F_{nat} \times B} \right) \times \left(1 \pm \frac{\hbar}{2 \times \Delta t \times \Delta E} \right) \times \left(\frac{E_{photon}}{E_{nat}} \right)$$

This equation accounts for the spin and frequency of particles in a magnetic field, as well as their natural field strength and the energy associated with photon interactions. By introducing Planck's constant (h), the equation links classical magnetic field theory to the quantum scale (Heisenberg 1927).

The Quantum Resonance Compliance Coefficient (QRCC)

The QRCC builds on the TDE by incorporating the Heisenberg uncertainty principle and the relationship between photon energy and natural energy. These variables define how resonance behaves when quantized, highlighting that resonance isn't just a wave phenomenon, but also has particulate properties at the quantum level. This understanding allows for precise control over resonance through adjustments in frequency, spin, and magnetic field manipulation.

$$E_{\text{photon}} = h \cdot f_{\text{TR}}$$

Example Calculation

To illustrate, consider a resonant frequency of 144 kHz. Using the TDE and factoring in Planck's constant, we calculate the amplified magnetic field strength:

$$TR = \left(\frac{720 \times 144 \times 1.5}{100 \times 1} \right) \times \left(1 \pm \frac{6.626 \times 10^{-34}}{2 \times \Delta t \times \Delta E} \right) \times \left(\frac{3.87 \times 10^{-19}}{1.60 \times 10^{-19}} \right)$$

The final amplified field is calculated at approximately 2.67 Tesla, demonstrating the ability to amplify a natural magnetic field through quantum resonance mechanisms.

Conclusion

The Transmagnetic Resonance Field Theory (TRFT), combined with the Quantum Resonance Compliance Coefficient (QRCC), offers a unified model for understanding and manipulating resonance at both classical and quantum levels. By incorporating Planck's constant and the Heisenberg uncertainty principle, this framework demonstrates that resonance, like energy, is quantized. This provides a deeper understanding of electromagnetic fields, allowing for advanced applications in fields ranging from energy manipulation to quantum computing.

Figures

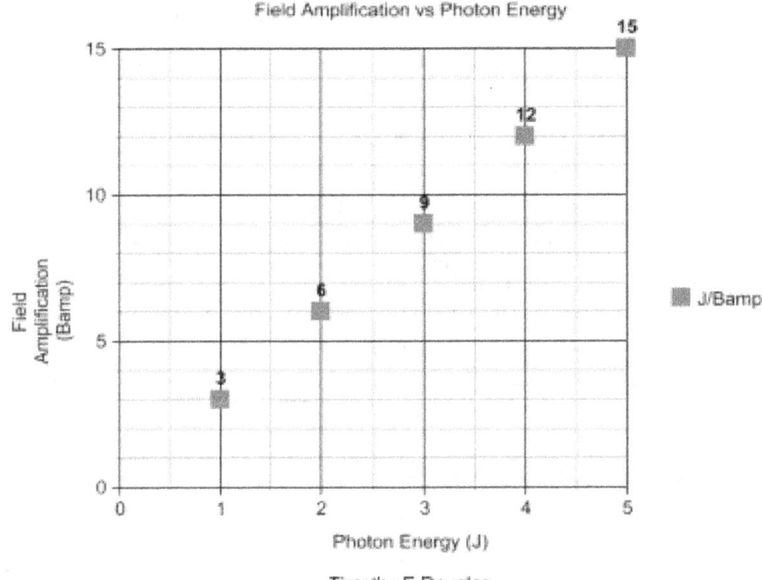

$$E_{\text{photon}} = h \cdot f_{\text{TR}}$$

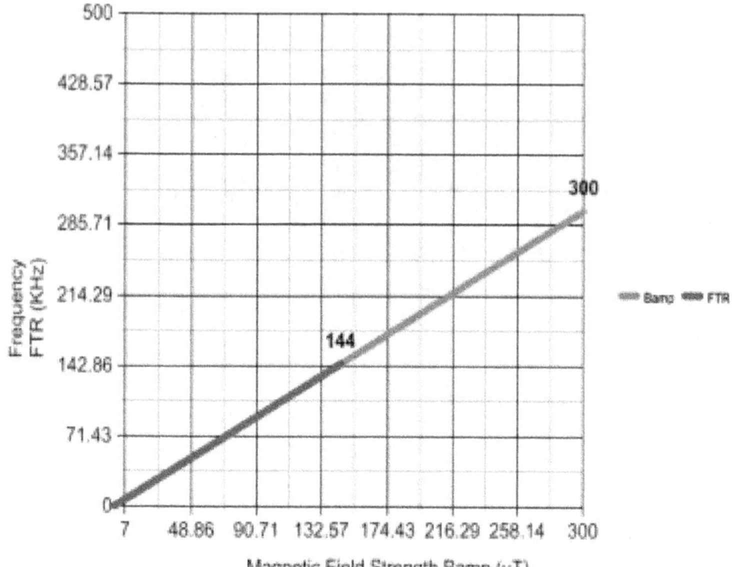

Timothy E Douglas

$$v_{\text{osc}} = \frac{720^\circ \cdot f_{\text{TR}}}{f_{\text{nat}}} \cdot \frac{E}{C_{\text{med}}}$$

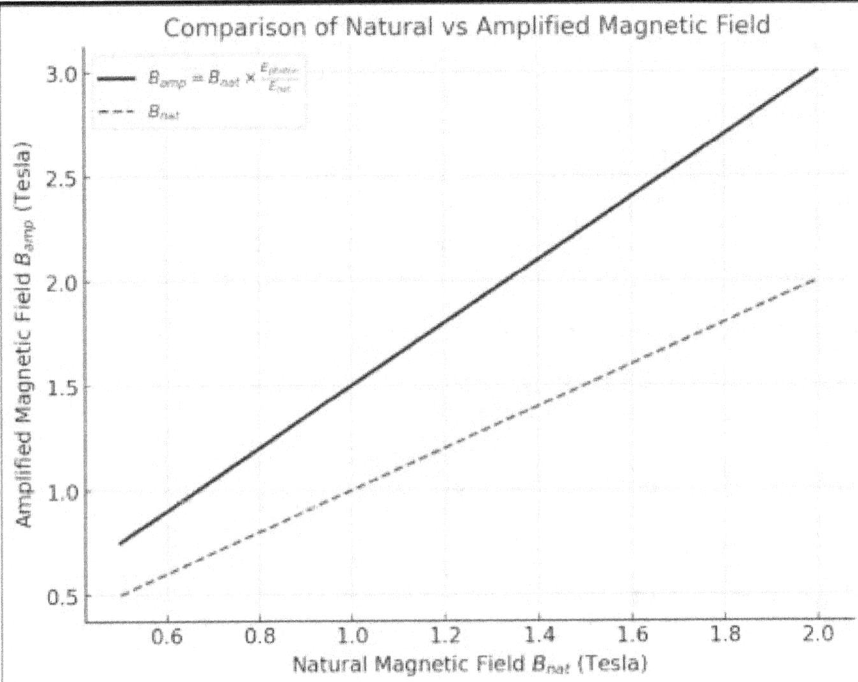

Timothy E Douglas

$$\text{Master Equation for Transmagnetic Resonance} = \left(\frac{720° \cdot f_{tr} \cdot \mu}{f_{nat} \cdot B} \right) \cdot \left(1 \pm \frac{\hbar}{2 \cdot \Delta t \cdot \Delta E} \right) \cdot \left(\frac{E_{photon}}{E_{nat}} \right)$$

Tensor Dynamics Equation:

$$v_{osc} = \frac{720° \cdot f_{TR}}{f_{nat}} \cdot \frac{E}{C_{med}}$$

Quantum Resonance Compliance Coefficient:

$$E_{photon} = h \cdot f_{TR}$$

Spin Coefficient:

$$\mu = g \cdot \frac{q \cdot \hbar}{2m}$$

Heisenberg Uncertainty Contribution:

$$1 \pm \frac{\hbar}{2 \cdot \Delta t \cdot \Delta E}$$

Magnetic Amplification Resonator Equation:

$$B_{amp} = B_{nat} \times \left(\frac{E_{photon}}{E_{nat}} \right)$$

The Transresonator Device:

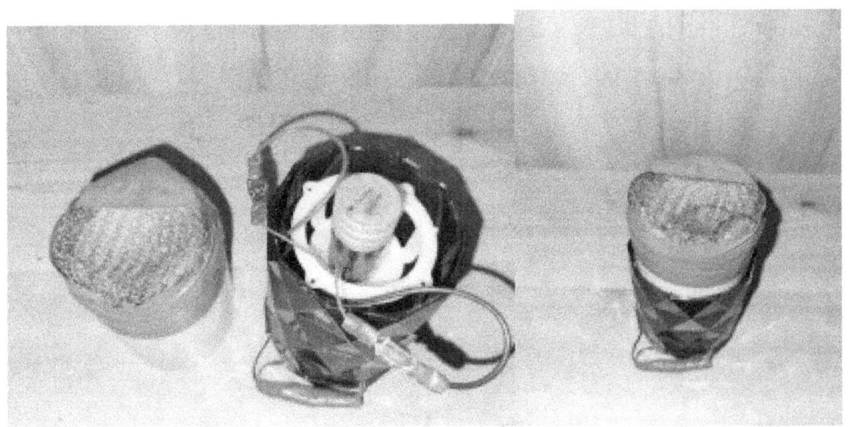

These pictures showcase the Transresonator device itself. Essentially what I did was affix natural magnets to an electromagnet, then attached the leads from the electromagnet to that of the amplifier module inside the bluetooth speaker. The copper netting and electrostatic tape are used as a Faraday Cap to weaken the base field which is approximately 22.1uT in the current version.

The first picture showcases the transresonator off and the base EM field approximately .42uT and the second picture showcasing the signal generator on at 144kHz, the speaker system off and the EM Field nearly doubling to .81uT.

Part 4

Resonance Matrix Hypothesis

Abstract

This paper introduces the Resonance Matrix Hypothesis, which explores the intersectional dynamics of multiple frequencies and their resonance interactions within a quantum system. By conceptualizing resonance as an array or matrix of electromagnetic wave lines, the hypothesis examines how overlapping frequencies create specific nodes of resonance exchange, where energy transitions and quantum leaps may occur. This model provides a mathematical framework that connects the number of intersecting points with increased resonance potential, suggesting exponential growth in interaction opportunities with the addition of each frequency. Incorporating the Heisenberg Uncertainty Principle, the hypothesis accounts for the inherent unpredictability in quantum behavior, especially concerning when and where resonant energy exchanges might occur. The implications of this hypothesis extend into both theoretical and experimental realms, offering new insights into the nature of quantum resonance, energy exchange, and the potential for broader applications in quantum mechanics and resonance-based technologies.

Introduction

The Resonance Matrix Hypothesis presents a pivotal framework for understanding the complex interactions between multiple frequencies and their corresponding resonances within a quantum system. This hypothesis posits that resonance can be conceptualized as a matrix or array formed by overlapping frequency lines, each representing distinct electromagnetic waves, akin to the primary colors of light—red, blue, green, and additional frequencies such as purple.

In this model, the overlapping or intersection points of these frequency lines serve as potential nodes for resonance exchange, wherein energy transitions can occur. As these lines travel in tandem, any twist, bend, or overlap creates opportunities for resonance interactions at these critical junctures. The probability of energy exchange at these nodes is influenced by various factors, including the inherent uncertainties described by the Heisenberg Uncertainty Principle.

The intersection points can be represented mathematically, with the number of potential resonance exchanges increasing exponentially with each additional frequency line introduced into the system. For instance, with three frequency lines, there are six points of resonance exchange; with four lines, there are twelve, and so forth. This exponential growth in interaction points suggests that complex energy exchanges may occur at these intersections, leading to significant quantum phenomena, such as electron transitions and shifts in energy states.

By incorporating the principles of quantum mechanics and the uncertainty inherent in particle interactions, this hypothesis aims to provide a comprehensive understanding of how resonance operates within a multi-frequency framework, laying the groundwork for further exploration and experimental validation in quantum resonance phenomena.

<u>Resonance as a Matrix Implications</u>

Intersection points: These are essentially where constructive interference happens. Each time the lines overlap, their waves can combine, amplifying each other (resonance), and could be identified as potential quantum "leaps." This concept ties in with the quantum mechanical view of energy transitions.

Moreover, adding more frequencies creates more points of interaction. When two waves of different frequencies overlap, depending on their phase relationship, you get either constructive or destructive interference. In the case of constructive interference, resonance happens, and the energy exchange is amplified.

When adding a fourth frequency, you start forming a 4-dimensional array of interaction points. The resonance matrix becomes more complex, especially with each added frequency, the number of points for resonance grows non-linearly.

Quantum Leap Hypothesis:

Quantum leaps happen when electrons jump between discrete energy levels. In this case, it's the intersection of the resonance matrix that might provide the energy required for such leaps. The exchange of electrons happens at these nodal points where resonance is strongest.

This could tie into quantum field theory, where particles and fields interact in similar ways—waves superimpose, and their interactions dictate energy exchanges.

Nonlinear Growth of Resonance Points:

This nonlinear increase with additional frequencies might correspond to higher-order harmonics in wave mechanics. These points of overlap could be analogous to harmonic oscillators in quantum systems, where higher resonances correspond to higher energy states.

If my idea can extend to quantum systems or electromagnetism, this model might help explain how wave interference patterns could lead to constructive resonance, influencing energy transitions in particles, fields, or even macroscopic systems.

<u>Concept</u>

Let's assume that each wave (denoted by different colours like Red [R], Blue [B], Green [G], and Purple [P]) represents a frequency or energy level in the system. These waves travel together and form constructive interference at specific overlap points, where energy amplification occurs, similar to resonance phenomena seen in classical and quantum mechanics.

Each time a new frequency component is added to the system, the number of intersection points increases exponentially, providing more opportunities for resonance exchange. In the quantum realm, this could explain the energy jumps or transitions (quantum leaps) where electrons move between discrete energy levels.

Defining Resonance Points:

Consider a system with N frequency waves interacting together. The intersection points for resonance exchange between these waves can be thought of as nodal points in a resonance matrix.

Equation for Resonance Points:

We can define a general form for calculating the number of resonance interaction points (RIP) based on the number of frequency waves present in the system.

$$RIP(N) = kN(N - 1)$$

Where:

- N = Number of frequency components (waves)
- k = A proportionality constant depending on the distance between waves and their phase relationship

This equation expresses that as you add more frequency waves, the number of resonance points grows quadratically. When these waves overlap, they produce resonance effects that amplify the energy exchange, potentially causing quantum leaps.

Example:

Let's take this scenario with four waves (R, B, G, P). For N = 4:

$$RIP(4) = k \times 4 \times (4 - 1) = 12k$$

Thus, in this scenario, the system would have 12 resonance points, where energy exchange could happen.

Incorporating Photon Energy:

At each of these resonance points, we can introduce the energy of a photon (Ephoton) to describe the quantum leap effect. The energy required to cause a quantum leap between resonance points can be derived from the relationship between photon energy and frequency, given by Planck's equation:

$$E_{\text{photon}} = h\nu$$

Where:

h = Planck's constant

ν = Frequency of the photon

Thus, each resonance point could be described as a point where the system absorbs or emits a photon, causing an electron to jump between energy levels.

Energy Exchange Formula:

Incorporating the energy of photons and the resonance points into the matrix hypothesis, it is proposed that the energy exchange at each resonance point can be described by:

$$E_{\text{res}}(N) = \sum_{i=1}^{N} E_{\text{photon}}^{i} \cdot RIP(N)$$

Where:

- E_res(N) = Total energy exchanged through resonance at N points
- Ephotoni = Energy of the photon involved at the ith resonance point
- RIP(N)= Number of resonance interaction points for N frequency components

This equation summarizes the total energy exchange across all resonance points in the system, with each photon contributing to the energy leap at a given resonance point.

Quantum Leap at Resonance Points:

The points of intersection where the frequencies overlap represent quantum leaps in the system, where particles or energy can "jump" between different states. This aligns with the idea that when enough energy is absorbed at a resonance point (i.e., when E_res(N) reaches a certain threshold), electrons or other quantum particles can transition between discrete energy levels.

Incorporating the Uncertainty Principle:

We could address the Uncertainty Principle by considering it in two ways: Macroscopic Resonance Points Are Certain, Quantum Transitions Are Not:

It could be argued that while the resonance points in the matrix are fixed (in terms of where waves overlap), the actual quantum leap or energy exchange at these points might be uncertain. In this case, the exact energy transferred or the timing of the transition could follow uncertainty principles. This would add a probabilistic element to when and how quantum leaps occur at the resonance points.

Furthermore, this principle suggests that there is an inherent uncertainty in energy (ΔE) and time (Δt) when discussing quantum transitions. So, while the resonance points themselves might be calculable, the exact amount of energy transferred and the time of transition would have some uncertainty.

While the resonance matrix describes the interaction points of multiple frequency waves, yielding fixed points of resonance exchange at the macroscopic scale, quantum mechanics introduces a layer of uncertainty at the subatomic level. Specifically, the Heisenberg Uncertainty Principle dictates that we cannot simultaneously know both the exact energy of a quantum particle and the precise time of its transition during resonance:

$$\Delta E \cdot \Delta t \geq \frac{h}{4\pi}$$

In this context, while the resonance points of interaction between waves are fixed, the exact timing and energy of the quantum leap at each resonance point remains uncertain. Therefore, each resonance point can be viewed as a probability node, where the quantum leap occurs probabilistically, aligning with the probabilistic nature of quantum states.

This quantum uncertainty emphasizes that while the resonance matrix provides a useful framework for predicting potential points of interaction, the actual energy exchange and transitions are subject to the inherent uncertainty of quantum mechanics.

Master Equation for the Resonance Matrix:

$$T(E, f) = \sum_{i=1}^{N_f} \left(\frac{h \cdot f_i}{1 + \Delta E_i \cdot \Delta t_i} \right) \cdot N_r(f)$$

Where:
- $T(E,f)$ is the total resonance matrix, which represents the overall quantum energy exchange at each resonance point.
- Nf is the number of frequencies involved in the matrix.
- fi is the individual frequency of each wave (for i-th frequency).
- h is Planck's constant.
- ΔE_i is the uncertainty in energy at the i-th resonance point (due to quantum fluctuations, from Heisenberg's Uncertainty Principle).
- Δt_i is the uncertainty in time for the i-th quantum leap.
- $N_r(f)$ is the number of resonance points, which can be given as:

$$N_r(f) = N_f \cdot (N_f - 1)$$

This term represents the number of potential resonance points created by overlapping frequencies.

Explanation of Terms:

1. Energy Exchange (E): Each frequency(fi) contributes a specific photon energy ($h \cdot f_i$), which is the energy exchanged at that frequency.

2. Quantum Uncertainty: The ratio $\dfrac{1}{1 + \Delta E_i \cdot \Delta t_i}$ represents the quantum uncertainty at each resonance point, where the Uncertainty Principle limits the precision of energy exchange at any given moment.

3. Resonance Points (N_r): This factor counts the number of resonance points based on the number of frequencies (N_f). Each pair of frequencies contributes to additional intersections and therefore more energy exchange possibilities.

Key Considerations:
- Planck's Constant (h) governs the energy of the photons at each frequency, ensuring the quantum nature of the exchange.
- The Uncertainty Principle limits our ability to know both the energy and time of the quantum leap with precision, represented by:

$$\Delta E_i \cdot \Delta t_i \geq \frac{h}{4\pi}$$

- The total number of resonance points grows as the frequency matrix expands with more overlapping frequencies, increasing the complexity of the system.

Example:
For three frequencies (R, B, G), this equation becomes:

$$T(E, f) = \sum_{i=1}^{3} \left(\frac{h \cdot f_i}{1 + \Delta E_i \cdot \Delta t_i} \right) \cdot (3 \cdot 2)$$

This represents how energy from each frequency is exchanged at 6 resonance points with the influence of quantum uncertainty applied to each.

This master equation encapsulates the total energy exchange in the resonance matrix by accounting for the quantum uncertainties and the increasing number of resonance nodes as more frequencies overlap. We can tweak the uncertainty factors and the number of resonance points depending on how many frequencies are involved and their respective quantum fluctuations.

Visualization of Resonance Matrix:
Consider visualizing this hypothesis as a matrix of intersecting wave frequencies, where each intersection point corresponds to a node of resonance. The more waves you add, the denser the matrix becomes, and

the more points of intersection appear. These nodal points are the places where energy exchange is strongest, similar to the nodes in a standing wave.

For example:
- With 2 frequencies (R and B), you have 2 intersection points.
- With 3 frequencies (R, B, G), you have 6 intersection points.
- With 4 frequencies (R, B, G, P), you have 12 intersection points, and so on.
This visualization aligns with how multi-wave systems like light or sound behave in resonance, creating amplified energy at nodal points.
Potential Implications:
1. Quantum Systems: In quantum mechanics, this matrix could represent different quantum states where energy is transferred via photons or wave interactions. Resonance exchange points might correspond to electron transitions, tunnelling events, or other quantum behaviours.
2. Electromagnetic Systems: In systems involving electromagnetic waves, such as your transresonator, these resonance points could represent areas where the magnetic and electric fields amplify, enhancing the overall energy of the system.
3. Applications in Technology: If properly harnessed, this matrix of resonance could be applied to energy systems, quantum computing, or field manipulation devices like the Transresonator built for these experiments. Understanding and controlling these resonance points might allow for more efficient energy transfer or manipulation of quantum states.

Next Steps for Exploration:
Empirical Testing: This hypothesis can be tested by constructing physical models where different frequency components interact and observing the resulting resonance points. Measuring the energy output at these points will validate the relationship between resonance and energy exchange.
Mathematical Validation: Further exploration of matrix theory and wave mechanics can help formalize this hypothesis. Matrix multiplication

techniques or eigenvalue analysis could be used to determine the precise number of resonance points in complex systems.

Quantum Field Application: Expanding this idea into quantum field theory, where fields interact in similar ways, could open up new areas of research in quantum resonance and energy manipulation.

Conclusion

In summary, this hypothesis presents an elegant way of visualizing how multiple frequency waves interact to create resonance, with the potential for quantum leaps at points of intersection. By formalizing this with equations and exploring it experimentally, we could unlock new understanding in fields ranging from quantum mechanics to electromagnetic energy systems.

Part 5

Pioneers of Quantum Mechanics

The 20th century saw an unprecedented revolution in the world of physics, particularly with the advent of quantum mechanics, which shattered classical notions of how the universe functions at its most fundamental levels. Where Newtonian mechanics once dominated, explaining the behavior of macroscopic objects, quantum mechanics emerged to explain phenomena on the atomic and subatomic scales—where particles behaved in strange, non-intuitive ways.

At the heart of this revolution were a group of brilliant physicists whose contributions redefined the scientific landscape. Their work introduced a radically new way of understanding the behavior of particles, energy, and the very nature of reality. These pioneers—Werner Heisenberg, Max Planck, Wolfgang Pauli, and Erwin Schrödinger—made indispensable contributions that form the foundation of quantum mechanics, and their insights continue to inform and inspire modern physics, particularly in fields like quantum resonance and magnetic field amplification.

This chapter will provide a deeper look at these visionary scientists, outlining their key discoveries and theories, and how those ideas underpin the broader concepts explored in this work. As we move through the various components of quantum theory, it becomes evident that the legacies of these thinkers are central to understanding the nuances of quantum resonance, wave mechanics, and the behavior of particles in energetic fields.

Werner Heisenberg (1901–1976)

Werner Heisenberg was a German theoretical physicist who, despite his relatively young age at the time, transformed the field of quantum mechanics with his groundbreaking Uncertainty Principle in 1927. Heisenberg's principle fundamentally challenged the deterministic view of classical mechanics, which held that the precise properties of a particle, such as its position and momentum, could always be measured and predicted. The Uncertainty Principle shattered this notion, asserting that the more precisely one measures a particle's position, the less precisely its momentum can be known, and vice versa.

This uncertainty is not due to flaws in measurement tools but a fundamental feature of nature itself. In the microscopic realm, particles exist in a haze of probabilities, only taking on definite properties when observed. This principle led to a profound reevaluation of the nature of reality, where observers play a crucial role in shaping the properties of the systems they measure.

Beyond the Uncertainty Principle, Heisenberg also developed Matrix Mechanics, one of the first formulations of quantum mechanics. Matrix mechanics described quantum systems using arrays (matrices) to represent observable quantities, like position and momentum, and provided a rigorous mathematical framework for the strange behaviors observed at the quantum level. This work would later play a role in the development of quantum field theory and resonate with many of the ideas about energy exchange in quantum resonance.

Key Contribution: Heisenberg Uncertainty Principle

$$\Delta x \cdot \Delta p \geq \frac{\hbar}{2}$$

This inequality expresses the fundamental limits to the precision with which certain pairs of physical properties can be known. It remains a foundational concept for explaining the behavior of particles at the quantum scale.

Matrix Mechanics would later influence the study of quantum resonance, where the probabilities associated with particle states and their dynamic interactions become central to predicting resonance effects within fields.

Max Planck (1858–1947)

Max Planck is often heralded as the father of quantum theory. Before Planck, physicists believed that energy behaved in a continuous manner, much like waves. However, Planck's research into black-body radiation led to an astonishing realization: energy is quantized. His famous equation, $E = h \cdot \nu$, revealed that energy is emitted or absorbed in discrete packets, which he called quanta. This radical departure from classical physics marked the birth of quantum mechanics.

Planck's work was primarily motivated by the need to resolve the ultraviolet catastrophe—a failure of classical physics to accurately predict the energy emitted by black bodies at short wavelengths. Through his quantization theory, he explained that electromagnetic energy could only be emitted in integer multiples of a fundamental unit, determined by Planck's constant (h), one of the most important physical constants in quantum theory.

Though initially hesitant to embrace the full implications of his own work, Planck's discovery would later become one of the cornerstones of quantum mechanics. His constant, (h), not only plays a critical role in quantum resonance but also underpins the energy transformations in the quantum realm. The relationship between frequency and energy is central to understanding how resonance occurs and how electromagnetic fields interact with quantum systems.

Key Contribution: Planck's Constant and Quantized Energy

$$E = h \cdot \nu$$

This equation shows that the energy of a photon is directly proportional to its frequency, where (h) (Planck's constant) acts as the proportionality factor. This discovery revolutionized our understanding of light, energy, and matter on the smallest scales.

Planck's work provides the framework for understanding how electromagnetic fields, such as those in magnetic resonance, function at the quantum level, where energy is exchanged in discrete amounts—essential for the field amplification equations discussed later.

Wolfgang Pauli (1900–1958)

Wolfgang Pauli, a Swiss-Austrian physicist, made significant contributions to quantum mechanics and is most famous for his formulation of the Pauli Exclusion Principle. This principle states that no two fermions (particles like electrons, protons, and neutrons that have half-integer spin) can occupy the same quantum state simultaneously. This simple yet profound rule explains the structure of atoms, particularly the arrangement of electrons in distinct energy levels, which prevents them from collapsing into a single state.

Pauli's Exclusion Principle was essential in the development of quantum chemistry and solid-state physics, providing insight into why matter behaves the way it does at a macroscopic level. Without this principle, the periodic table, and indeed, much of our understanding of the physical world, would be impossible. This principle also has deep implications for the study of spin dynamics, one of Pauli's other major contributions.

In the context of quantum resonance, Pauli's work on spin theory is crucial. Spin describes a particle's intrinsic angular momentum and is key to understanding how particles interact with magnetic fields. The behavior of particles in quantum resonance is deeply tied to their spin states, and the exclusion principle helps explain why only certain states are allowed, creating the conditions necessary for resonance exchange to occur at certain energy levels.

Key Contribution: Pauli Exclusion Principle

States that no two fermions can share the same quantum state, an idea that governs the arrangement of electrons in atoms and plays a critical role in quantum mechanics.

Pauli's insights into spin dynamics and particle interactions remain essential for understanding how particles engage with fields and with each other, creating the conditions for quantum resonance.

<u>Erwin Schrödinger (1887–1961)</u>

Erwin Schrödinger was one of the central figures in the development of quantum mechanics. His greatest contribution is the Schrödinger Equation, which provides a mathematical description of how the quantum state of a system changes over time. His formulation of wave mechanics showed that particles could be described not just as discrete objects but also as waves, a revolutionary concept that led to a deeper understanding of the dual particle-wave nature of matter.

Schrödinger's equation, expressed in both time-dependent and time-independent forms, remains one of the most important equations in quantum mechanics. It allows physicists to calculate the probability of finding a particle in a given state and describes the evolution of systems as they interact with external forces, such as electromagnetic fields.

The Schrödinger Equation is pivotal to quantum resonance, where it can describe how particles behave within oscillating fields. By solving this equation, physicists can predict how energy is transferred between different states, which is essential for understanding both natural and amplified resonance effects in quantum systems.

Key Contribution: Schrödinger Equation

$$i\hbar\frac{\partial}{\partial t}\Psi(x,t) = \hat{H}\Psi(x,t)$$

This equation describes how the quantum state (represented by the wavefunction (Ψ) of a system evolves over time, where $\hat{H}$ represents the system's total energy (Hamiltonian).

Schrödinger's work is vital for predicting how quantum resonance behaves within magnetic fields, explaining how particles oscillate between energy states and how resonance points develop as fields overlap or interact.

<u>Nikola Tesla (1856–1943)</u>

Nikola Tesla, one of the most prolific and visionary inventors in history, revolutionized our understanding of electricity, magnetism, and resonance. Born in the Austrian Empire (modern-day Croatia), Tesla was a brilliant electrical engineer and physicist who is best known for his contributions to the development of alternating current (AC) power transmission. Tesla's groundbreaking ideas paved the way for many modern technologies, but his lesser-known work on resonance and energy transmission is where his true genius shines.

Tesla's work on resonant circuits and wireless energy transmission was decades ahead of his time. He demonstrated that electromagnetic waves, when tuned to specific frequencies, could be transmitted without the need for wires—a principle he sought to harness with his famous Wardenclyffe Tower. This idea was rooted in his theory that energy could be transmitted efficiently by creating resonance between the transmitter and receiver. Tesla was particularly fascinated by the Earth's own resonant frequency, which he believed could be tapped into to transmit energy globally, essentially providing free energy to all.

Though many of his grander visions, such as global wireless power and large-scale resonant energy transmission, were never fully realized, Tesla's work continues to inspire modern researchers. His deep understanding of the relationship between frequency, energy, and resonance laid the foundation for many of today's cutting-edge technologies, including those explored in this book. Tesla's dream of using resonance to unlock new energy sources resonates with the fundamental principles of transresonance explored in the *Resonance Matrix Hypothesis* and the *Transresonator* device.

Tesla's legacy is one of boundless creativity and an unyielding belief that resonance and frequency hold the key to untapped energy, a vision that continues to guide new generations of inventors and researchers.

The contributions of Heisenberg, Planck, Pauli, Tesla and Schrödinger form the bedrock of modern quantum theory. Their work on uncertainty, quantization, exclusion, and wave mechanics not only revolutionized our understanding of physics but also set the stage for much of the research into quantum resonance, magnetic fields, and energy transfer. Understanding the theories these pioneers developed allows for a more nuanced exploration of resonance and field mechanics in modern applications, especially within the frameworks laid out in this work.

Works Cited

(1) Werner Heisenberg. The Physical Principles of the Quantum Theory. University of Chicago Press, 1930.
This is a foundational work where Heisenberg discusses the principles of quantum mechanics, including the uncertainty principle.
(2) David C. Cassidy. Uncertainty: The Life and Science of Werner Heisenberg. W.H. Freeman & Company, 1992.
A biography of Heisenberg that explains the historical development of the uncertainty principle in the context of his life and work.
(3) Werner Heisenberg. Physics and Philosophy: The Revolution in Modern Science. Harper & Row, 1958.
Heisenberg reflects on the philosophical implications of quantum theory and the uncertainty principle in this accessible work.
(4) Max Planck. The Theory of Heat Radiation. P. Blakiston's Son & Co., 1914.
This is Planck's seminal work, where he introduces Planck's constant in the context of black-body radiation.
(5) Helge Kragh. Quantum Generations: A History of Physics in the Twentieth Century. Princeton University Press, 1999.
Offers an in-depth look at the historical context and significance of Planck's contributions to physics.
(6) Max Planck. Eight Lectures on Theoretical Physics. Columbia University Press, 1915.
In this collection of lectures, Planck elaborates on quantum theory and his work with Planck's constant.
(7) Malcolm H. Levitt. Spin Dynamics: Basics of Nuclear Magnetic Resonance. John Wiley & Sons, 2001.
A comprehensive guide on spin dynamics in NMR, covering both the theoretical and practical aspects.
(8) Daniel C. Mattis. The Theory of Magnetism. Springer, 1981.

Focuses on the theoretical aspects of spin dynamics and magnetism, providing a detailed examination of spin-spin interactions.
(9) Charles P. Slichter. Principles of Magnetic Resonance. Springer, 1990.
A thorough introduction to the principles of magnetic resonance, including a focus on spin and its role in NMR.

(10)Wigner, E. P. (1959). Group Theory and Its Application to the Quantum Mechanics of Atomic Spectra. Academic Press. This work explains the role of matrices in quantum mechanics and their application in atomic spectra analysis.
(11)Heisenberg, W. (1925). Quantum-theoretical Re-interpretation of Kinematic and Mechanical Relations. Zeitschrift für Physik, 33, 879–893.
Heisenberg's paper on matrix mechanics, the foundation of quantum mechanics.
(12)Dirac, P. A. M. (1930). The Principles of Quantum Mechanics. Oxford University Press.
One of the early authoritative texts on quantum mechanics, discussing matrix theory extensively.

(13)Schrödinger, E. (1926). Quantization as an Eigenvalue Problem. Annalen der Physik, 79, 361–376.
Schrödinger's fundamental work on wave mechanics and the wave equation.
(14)Debye, P. (1926). Quantum Theory and the Equivalence of Wave Mechanics and Matrix Mechanics. Zeitschrift für Physik, 34, 681–685.
Discusses the conceptual bridge between wave and matrix mechanics in quantum theory.
(15)Maxwell, J. C. (1865). A Dynamical Theory of the Electromagnetic Field. Philosophical Transactions of the Royal Society of London, 155,

459-512. Maxwell's equations form the foundation of electromagnetism, crucial for understanding magnetic field behavior.

(16)Feynman, R. P., Leighton, R. B., & Sands, M. (2011). The Feynman Lectures on Physics, Vol. II: Mainly Electromagnetism and Matter. Basic Books. This book covers the principles of electromagnetism, including discussions on magnetic fields and their amplification.

(17)Jackson, J. D. (1998). Classical Electrodynamics. Wiley. A comprehensive resource on classical electrodynamics, covering topics related to magnetic fields, their amplification, and applications.

About the Author

Timothy E Douglas is a researcher, theorist, and creator of the Transresonator, a groundbreaking device exploring the interaction of quantum resonance with electromagnetic fields. With a background in psychology, ehtics, and an autodidactic journey through quantum mechanics, Timothy has developed novel concepts at the intersection of physics, technology, and metaphysical exploration. Known for challenging traditional academic approaches, Timothy combines practical experimentation with a deep intuitive understanding of quantum phenomena, offering fresh perspectives that could change the way we understand energy and resonance. This is Timothys's debut publication into Quantum Physics, part of a larger project aimed at unlocking new potentials in energy research and human understanding.